SCARY MONSTERS AND
BRIGHT IDEAS

ROBYN WILLIAMS is a broadcaster who has presented 'The Science Show' on Radio National for 25 years. He is a Fellow of the Australian Academy of Science, and of Balliol College, Oxford, and a visiting Professor at the University of New South Wales.

Other Frontlines

A Bill of Rights for Australia George Williams

Australia's Economic Revolution John Edwards

Black Lives, Government Lies Rosalind Kidd

Deadlock or Democracy: The Future of the Senate Brian Costar (ed),
with Meg Lees, Helen Coonan, John Faulkner and Harry Evans

Employer Controls Over Private Life Ronald McCallum

Gambling Government: The Economic and Social Impacts
Michael Walker

In Defence of Globalisation Keith Suter

Taxing Times: A Guide to Australia's Tax Debate John Quiggin

The Wik *Debate: Its Impact on Aborigines, Pastoralists and Miners*
Frank Brennan

FRONTLINES

ROBYN WILLIAMS

SCARY MONSTERS
AND BRIGHT IDEAS

UNSW
PRESS

To the memory of Ian Anderson, Australian editor of
New Scientist. A great journalist and a dear friend.

A UNSW Press book

Published by
University of New South Wales Press Ltd
UNSW SYDNEY NSW 2052
AUSTRALIA
www.unswpress.com.au

© Robyn Williams 2000
First published 2000

National Library of Australia
Cataloguing-in-Publication entry:

  Williams, Robyn.
  Scary monsters and bright ideas.

  ISBN 0 86840 707 0

  1. Science and civilization. 2. Science —
  Social aspects — Australia. 3. Genetic
  engineering — Australia. 4. Research —
  Australia — Finance. I. Title. (Series:
  Frontlines)

  509.94

Printer Brown Prior Anderson

CONTENTS

Introduction   6

1 Dr Frankenstein and Dr Faust   8

2 Popularise or perish?   28

3 Inventing the future   47

Conclusion   60

INTRODUCTION

In April 2000 I was doing a tour of the Eye Hospital at the University of Melbourne. We left the main building with its long corridors, slow lifts and patched patients waiting for treatment and walked briefly in the autumn sun to a faded mansion with filigree ironwork now doing duty as a lab. Up the wide staircase, past rooms with machines going 'ping!' and into a crowded area that once may have been a salon for the gentlefolk of Carlton, where several young Australians were at their microscopes.

'This is the team working on the genetics of macular degeneration [a form of blindness],' announced the prof in the bow tie. They smiled politely.

'Where are you from?' I asked the nearest of them, feeling like Princess Margaret, queen of the meaningless question.

'I'm from law,' he replied. 'I'm doing this as part of my law degree. Twenty per cent of us at Melbourne are doing science within our legal qualification. Soon I expect it will be fifty percent!'

Birds sang. The sun shone. At last the year 2000 had started.

I have been a science journalist since 1972. In those days we were intrinsically sceptical about what Harold Wilson had called 'the white heat of the technological revolution'. We knew he was waffling about some magic box of promises because the floor had fallen out of the economy on the night he'd moved into Downing Street and he had to say something! We also knew that half of science was compromised by its role in war and the other half was suspect because it was in the hands of naifs — there had been too many lobotomies and Cuckoo's Nests, Silent Springs and Love Canals. Science was not about to save the world, but it could help. Scientism was the enemy. Remember Lysenko.

Time passed. Science itself proceeded apace. As it did so we, both the interpreters and the public we served, tried to keep up. What was bad one day became fine the next. Then the other way around. The products of scientific research became more and more profitable and the stakes higher. It was not surprising that debates about these forces came only intermittently — we are constantly concerned with jobs, mortgages and our heroes' groins, less so about 'whither society?'.

But science, throughout, remained someone else's property. Sadly, it never became 'ours'. That it should is not a preposterous suggestion. It is, after all, our beach, our grass, our body, our sky, and it all hangs together in a coherent way that is compellingly interesting. But that interest stayed at school, university and institution, within the edifice, not the village. Unlike art and local politics, scientific culture in Australia is still, despite occasional young lawyers doing genetics, overwhelmingly separate. It is 'their' science. And it suffers accordingly. I did not expect the healthy reserve I applauded thirty years ago to become the alienation I see so starkly today. It is to our cost.

These essays are about three elements of that alienation. The first is the 'scary monster' of genetic engineering, seen by many in research as, essentially, an anti-science movement. The second is on public awareness of science and the exquisitely thin line between being a reporter and leading the cheer squad. The third is about 'bright ideas' and policy, which really means money. Without being partisan about funding for R & D, I nonetheless find it extraordinary that a country like Australia is prepared to spend $105 billion per annum on gambling but cannot find money to study the long-term effects of bleaching on the Great Barrier Reef, one of its unique treasures.

I am guided by the thoughts of two writers from my youth: HG Wells, who remarked, 'Human history becomes more and more a race between education and catastrophe'; and Bertrand Russell, who noticed, 'The trouble with the world is that the stupid are cocksure and the intelligent are full of doubts'.

1 | DR FRANKENSTEIN AND DR FAUST

Mary Shelley was eighteen when she began writing *Frankenstein; Or, The Modern Prometheus*. She started the story in June 1816, as she says, after 'amusing ourselves with some German stories of ghosts ... [which] excited in us a playful desire of imitation'. The 'us' included the man she was to marry later that year, Percy Shelley, and his friend, Lord Byron. All three were familiar with the scientific ideas that had galvanised European society in the past decades and which were, especially in Britain, producing the Industrial Revolution.

Frankenstein's monster, both in that original (1818) text and in the classic film starring Boris Karloff, is not automatically a villain. He is misunderstood by those who encounter him, in part because of his grotesque appearance and unnatural power, but also because of his origins. In some quarters it was thought unseemly for human beings to be playing with some kind of life-force, surely the gift of God (or Prometheus?). Experiments by Galvani and Faraday with electricity, hinting at the basis for this life-force, conjectures by Humphry Davy and Erasmus Darwin on its evolutionary nature, gave the young woman plenty of material for her 'playful desire'.

The scheming scientist is revolted by his creation with its 'superhuman' powers and addresses it accordingly in the second part of the book:

'"Devil" I exclaimed "do you dare approach me? and do not you fear the fierce vengeance of my arm wreaked on your miserable head? Begone, vile insect! or rather stay, that I may trample you to dust!"'

Not the greetings of a proud daddy. Ian Wilmut is much kinder about Dolly. And Monsanto far more enthusiastic about its own GM progeny. So far.

'I expected this reception,' replies the monster (here called the daemon). 'All men hate the wretched; how then must I be hated, who am miserable beyond all living things! Yet you my creator, detest and spurn me, thy creature, to whom thou art bound by ties only dissoluble by the annihilation of one of us ... How dare you sport thus with life? Do your duty towards me, and I will do mine towards you and the rest of mankind.'

The monster is misunderstood. The scientist is a dissembler. One has the irresistible feeling on reading Mary Shelley's novel that the story could have been different if Dr Frankenstein had not been such a shit (and unable to propagate in the normal way) and if the spectacular creation of the new science had been better understood in the first place. As it was the pressures on the author were such that she had to rewrite the novel twice to appease establishment prejudice. The atheistic Shelleys had been, perhaps, too cavalier for the sensibilities of the early nineteenth century. The subtleties of the original ideas had to be reset for the pieties of 1831. The scholar Marilyn Butler of Exeter College, Oxford records how 'much pressure, direct and indirect, was put upon the author to change the book' and how much Mary Shelley and her publishers may have regretted the bowdlerisation.

It is therefore with decidedly mixed feelings that some of us respond to the words 'Frankenfoods', 'Frankenfish' and 'Frankenscience' which appeared in the British press in the late 1990s and now commonly appear in Australian newspapers as well. Are subeditors canvassing the full subtleties of Shelley's imaginative conception, with its questioning of the nature of life itself and, indeed, the unlikely role of a deity in its manipulation, or are they simply socking us some purple headline material? Are they responding to a third or fourth industrial revolution with all its consequences and uncertainties as were the young writers by that Swiss lake near Geneva, or are they merely lobbing scary buzzwords?

The most useful questions to ask, in examining the GM controversy in this context, are fourfold:

1   Is the debate being conducted in a manner likely to enlighten a concerned public, or to confuse it?

2   Is the *science* behind genetic modification best separated from the *finance*?

3   Is there a qualitative difference between invisible entities such as genes (and nuclear power — both potentially capable of producing monstrosities) and other environmental issues?

4   Could GM distract the community from major environmental affairs at a crucial time?

Let us consider these in turn.

Enlightenment?

In October 1999 headlines appeared in the British press announcing that Dr Arpad Pusztai, a 'world expert in lectins, the natural insecticides produced by plants' and perhaps the best known GM whistleblower, had been proved right. 'Smeared GM expert vindicated,' announced the *Independent On Sunday*, adding the breakout 'How I told the truth and was sacked'.

'Vilified gene food scientist right after all,' followed the *Sydney Morning Herald*. This was in response to the decision of the medical journal, the *Lancet*, to publish Dr Pusztai's paper claiming to show that rats fed certain GM potatoes had changed gut linings.

The Hungarian research scientist had been working at Aberdeen's Rowett Research Institute when he had been pressed on this work during a television interview in mid-1998. 'I was absolutely confident I wouldn't find anything. But the longer I spent on the experiments, the more uneasy I became,' he said. The reaction to this original 'warning' about the potatoes was immense and global. Some took it as the first official evidence that GM foods did harm. Many assumed that it was based on published, peer-reviewed research — it wasn't. Dr Pusztai's employers asked him to cease public comment and to take early retirement. Only fourteen months later was a formal paper considered adequate for release and then in the face of severe criticism from referees.

'Under other circumstances this work would probably not have been published. There are a number of unsupported statements, and much evidence of muddled thinking,' wrote Professor John Gatehouse, who examined the paper and sent his comments to the Royal Society. At the time of Dr Pusztai's original public comments the Royal Society had said it looked forward to the proper publication of the potato experiments. The question is: was the paper placed in the *Lancet* not because of its quality but its notoriety? Could it be that a venerable scientific journal would compromise its standards to accommodate a communication of dubious merit? Such was the consternation that the editor of the *Lancet*, Richard

Horton, chose to defend his decision in a two-page article in the *Times Literary Supplement* (17 December 1999):

> On August 10 1998, and with full support and encouragement of his scientific superiors ... Pusztai claimed on ITV's World In Action that genetically modified potatoes harmed laboratory rats. By doing so, he broke a fundamental scientific rule, an action that was widely condemned by his colleagues. The rule states that scientists must not report the results of their research to the public before those findings have been presented to their peers either at a scientific meeting or in the pages of a scholarly journal. It seems like a gagging clause, and it is. At one level, the rule is well meant: to allow other scientists to pass judgement — peer review — for or against a piece of work before letting it loose on the public. At another level, the rule imposes a tight constraint on what is deemed acceptable science.

Richard Horton feels this constraint puts 'a conservative brake on new ideas'. But he sees the importance of caution in medicine where health scares or radical approaches to therapy could be ephemeral and misguided. In Pusztai's case Horton writes that the geneticist perceived a gap between the research findings of his group and claims made by corporations and that 'commercial pressures might have prevented his work from being published'.

> The public storm that followed these events forced Pusztai to seek formal publication of his findings. In December 1998, he and a pathologist, Stanley Ewen, submitted a research letter ... to the Lancet. We sent their letter to six external advisers ... Four of these peer reviewers recommended publication after clarifications and revisions. Our nutritionist noted that the paper was 'seriously flawed'. Nonetheless, his judgement was that he 'would like to see [the work] published in the public domain so that fellow scientists can judge for themselves. If the paper is not published, it will be claimed there is a conspiracy to suppress information.' Our sixth reviewer, an agricultural geneticist, also argued that Pusztai's work was 'badly flawed' and that the 'letter should be rejected'.

Here was the tension. If Horton rejected the paper, on good grounds, he could be accused (and surely would, given the intense interest in both Pusztai and GMOs) of censorship against the public interest. If he published he would be pilloried for lowering standards, but the editor of the *Lancet* chose to publish for two reasons. One was to continue a tradition, going back he said to the early part of the seventeenth century, of courting controversy, of daring to speak out. The second was a scepticism of peer review which he says is 'often prejudiced, unjust, incomplete, sycophantic, insulting, ignorant, foolish and wrong'. The real test of an experiment, he says, is not opinion but repeatability. He cites the rejection in 1797 by the Royal Society's *Philosophical Transactions* of Edward Jenner's work on vaccination and smallpox as an extreme example of the dead hand of caution.

But Horton's accusations against the Royal Society and the scientific establishment went further. He accused it of being seriously compromised by its association with business.

> Public-private partnerships with industry are being established in all spheres of the Royal Society's work. There are twenty-one industry-sponsored fellowships, and the Royal Society notes in its latest annual report that 'links between industry and academia are being strengthened' through these schemes. Some examples: the RS/Glaxo Wellcome research professorship, the Rolls-Royce plc sponsored industrial fellow, the Zeneca lecture, the Richard Branson innovation lecture, RS/Esso grants, a RS/Esso energy award, a Glaxo–Wellcome medal, and additional major donations from BP Amoco, GEC, Esso UK, Glaxo Wellcome, National Westminster Bank, Rhone Poulenc UK, Rolls Royce, Filtronic, Nycomed Amersham, Allied Materials Europe and Zeneca.

All these names are trotted out and repeated by Horton — and implied to influence the Society's activities. 'This admission sits oddly,' he writes, 'with the Society's claims to "its unique combination of scientific authority and independence from vested interest".'

Anyone who has been responsible for a research institution in the last few years (I myself was President of the Australian Museum Trust), a period marked by massive cutbacks in public funding, knows how much one is forced to turn to the private sector. The alternative is to cut programs or to shut up shop altogether. The presence, therefore, of corporate names on Royal Society (or Australian Academy) plaques is hardly surprising. But how does the Society itself respond to Richard Horton's indictment? Has science been compromised by its Faustian connections to become transmogrified into a modern Frankenstein?

Nobel Laureate and President of the Royal Society, Sir Aaron Klug, wrote to the *Times Literary Supplement* in reply. His letter was not printed until 31 December 1999! Here are some extracts:

> We became involved in the GM debate, not because we have a particular mission to defend the interests of biotechnology, still less because of vested financial interests ... but because it is the Society's duty to see that, where appropriate, public policy should be based on the best available science rather than propaganda and emotion. The assertion that we have taken on 'the role of public relations to support the genetically modified food industry' is false.
>
> The second charge is ... that connections with corporate donors have compromised [our] independence. This is simply not true. The holders of industry-sponsored RS Fellowships or Professorships are not chosen by the donors, but by independent panels ... nor do any such donations affect what policy issues we take up or what we conclude.

Sir Aaron Klug, about to hand over his presidency of the Royal Society to the Australian Sir Robert May, concludes acerbically:

> The Society is not for sale, and none of our many donors imagines for a moment that it is ... To write, as Horton does, that the Society has 'to keep its business investors happy' is as ludicrous as it is offensive.

It is worth remembering that all this vituperation is over one paper by a previously unremarked scientist concerning the ambiguous effects of unusual potatoes on a very few rats. No bombs, no deaths, no lobotomies, no felonies, no frauds — none of the issues that had exercised scientists in the public arena during the preceding turbulent century. Yet the fuss over GM organisms was eclipsing everything else, and not only in Britain.

Peter Raven has been director of the renowned Missouri Botanical Garden for over thirty years. He is to plants what Paul Ehrlich is to animals: a conservationist of immense stature and achievement. This much was duly recited in a six-page portrait of him in the *Riverfront Times*, one of St Louis' best-respected newspapers. Then it added that he had attracted 'the confusion of environmentalists who can't decide whether one of their heroes has been bought'.

This line is put bluntly in the subheading to the article — main title: Peter and The Wolf — 'Why Missouri Botanical Garden's Peter Raven, world-renowned environmentalist, courts Monsanto's favour, boosts its biotech and takes its money'. Monsanto is the 'Wolf' in this parable.

The 'case' against Dr Raven is circumstantial. The Gardens received US$3 million in their last fundraising campaign (according to the newspaper) from Monsanto. They also contributed much of the US$146 million given to set up the Danforth Plant Science Centre and had their name given to the Monsanto Centre — both at the Gardens. In addition to all this, Dr Raven's new wife is Monsanto's head of public policy!

But what has Raven actually *done*? He clearly has connections with the 'Wolf' but does that mean he must have fleas? Nowadays, that is assumed, especially with the dreaded Monsanto. Raven is quoted in reply as follows:

> It ought to almost be a moral question whether people start labelling people because they're married. I have been a very well-regarded scientist for all my career and achieved everything I wanted to achieve, more than most, and I didn't do that because I was married. Kate, if anything, is more of an environmentalist than I am. She works with public acceptance, listening and bringing people together and trying to find ways Monsanto can operate effectively and well with public approval. Her job is not selling or convincing people of anything.

Raven's guilt is by association — Monsanto is a bruiser and some of us have experienced what appeared to be their bullying tactics. They have had GM crops on the market for nearly five years but already (according to *Riverfront Times*) their products cover 52 million acres in the United States alone. But is there a real skeleton in this case? Has Raven gone beyond enterprising fundraising by lending his good name as a conservationist to a company with the reputation of a carnivore? The nearest I could get was a statement from one or two fellow academicians saying he had been 'naive'.

This 'naivety' does not seem, in the end, to have done him much harm. In March 2000, Dr Peter Raven was elected the next President (from 2001) of the American Association for the Advancement of Science (AAAS).

These are two of the best known examples of GM controversy. If they demonstrate anything in common it is first, that the public is highly suspicious of the motives of powerful corporations; secondly, that scientific bodies, however noble, are also regarded charily. As Sir Robert May told the AAAS meeting in Philadelphia in 1999, 'if you imagine that public opinion will be placated by half a dozen men in suits saying "trust me I'm a scientist" — then forget it. That era is over!'

What about the science?

Genes code for enzymes. Each enzyme mediates a chemical reaction. Few processes in an organism are governed by only one gene. That is why the talk about 'genes for ...' gayness, religiosity, schizophrenia, even alcoholism, are so misleading. It is also preposterous to isolate genes as if they belong only to one species: 'The rat that thinks it's a Brussels sprout'; 'The tomato that's part fish'; and so on.

Scientists will say that they are modifying an organism minimally, perhaps to add pest resistance (the essence of Dr Pusztai's potato). They point out that longer sections of the genome are transferred during old-fashioned cross-breeding. Critics claim this is ingenuous. They say that the gene's potential is changed by its new surroundings. This is the thrust of David Suzuki's concern and he reminds us of it in his 1999 book, *Naked Ape to Superspecies*, and that he himself is a former professor of genetics and should know: 'In their obsession with specific goals and fascination with their technical dexterity, scientists seldom reflect seriously on the social or environmental ramifications of their work.'

Critics of GMOs also warn that the altered organisms, mostly plants, will have an uncertain effect, over the indefinite future. Genes let loose in the novel strains may themselves leak into other species with unpredictable consequences. Could pest-resistant crops give rise to pest-resistant, even herbicide-resistant weeds? Will triffids march?

Dr Alan Kerr tried to answer some of these questions in a talk he gave on Radio National. He is one of the many scientists bewildered by the onslaught. He had seen himself as a responsible professional responding, in a way, to Rachel Carson's warning in 1962 of *Silent Spring*. His professional life was dedicated to the development of 'biological' alternatives to chemicals and this was celebrated ten years ago by his receipt of the very first Australia Prize for his work on crown gall disease, a form of plant cancer that can be exploited for genetic engineering purposes.

In the talk Dr Kerr asked: Are genetically engineered plants a danger to the environment? Will herbicide resistance spread? What are the advantages and disadvantages of the crops? Are genetically engineered crops toxic? Should there be a moratorium on genetic engineering? Here are some of his answers:

1   Are GMOs an environmental danger?

'Most of the criticism is directed against herbicide-resistant crops. I freely admit that the availability of herbicide-resistant crops will lead to a greater use of herbicides. Is this a bad thing? In some ways, yes. Nobody wants to spread around more toxic chemicals unless there is a very good reason to do so ... The only real alternative in Australia, is soil cultivation. In many soils this leads to erosion. The great beauty of herbicides is that soil cultivation can be eliminated or reduced to a minimum, thus preventing soil erosion, which, along with salt damage, is the greatest environmental problem in Australia.'

2   Could herbicide resistance spread?

'This is a good point and it could happen. In Australia, there are very few crops that inter-breed with weeds. An exception is canola ... However, the likelihood of this happening can be tested experimentally and an assessment committee like GMAC [Genetic Manipulation Advisory Committee] is well aware of such hazards. If the dangers outweigh the benefits, the resistant crop will not be grown.'

3   What are the advantages of GM crops?

Insect-resistant crops can 'reduce the use of insecticide by half ... In America, it has reduced insecticide use on cotton by about 90 per cent.'

4   Are genetically engineered crops toxic?

'Naturally, if you put a toxin-producing gene into a plant consumed by humans, the plant will be toxic. But who would want to do that?'

5   Should there be a moratorium for five years to prove there is no deleterious effect to humans or the environment?

'This is mere humbug. Everyone knows, or should know, that scientifically, you cannot prove a negative. You can show the probability is very low but that will not satisfy the critics. I sense it is more than opposition to genetic engineering. It is an anti-science movement.'

This is the nub. And it could be where Dr Frankenstein's monster really appears. Dr Kerr goes on:

It is interesting to listen to discussions on genetically engineered crops. Almost inevitably, three points will be raised: 1. mad cow disease in Britain; 2. the Chernobyl disaster; and 3. the dioxin contamination of chicken feed in Belgium. None of these has anything to do with genetic engineering.

Except, Dr Kerr, for two things: harassed politicians wanting to hose down public concern; and an industry, reluctant to spend large amounts on ensuring public safety, telling porkies to the bitter end. But Dr Kerr's bewilderment, as a lifelong environmentalist, is understandable. After all, it was the scientists themselves, when recombinant DNA technology was first announced in the mid-1970s, who called for and applied a moratorium on its use. During this moratorium, the academics and citizens of Cambridge, Massachusetts, led a remarkable exercise in public debate in which the issues were canvassed for a year, after which the geneticists went back to their benches. No one had asked them to do otherwise.

Will there be a moratorium in Australia? Already we have seen GM crops being attacked physically. In March 2000 some pineapples (which had been altered by the reintroduction of another type of pineapple gene!) were dug up and vandalised by activists. But there has been no vandalism on the scale of that in Germany or Britain. Dr Jim Peacock, chief of the CSIRO Division of Plant Industry says the research will continue:

> CSIRO is not considering 'holding off' on releasing GM cereals. But we believe the commercial release of modified cereals may be as much as five years away, because it will take this time to develop and test new varieties and ensure all material is rigorously assessed by Australia's regulatory organisations. In the next five years we will have better crop varieties which use the information gained from gene technology research.

Dr Peacock is, significantly, one of those fellows of the Australian Academy of Science keen to bring awareness of genetics to the public. He was the prime mover in producing a video called 'Can Pigs Fly?' presented by science reporter, Dr Gael Jennings, shown in supermarkets and other public venues all over Australia in the 1980s. He has also built a fabulous 'Discovery Centre', attached to the Division's laboratories, that allows the public close-up contact with the botanists doing the gene work and the chance to interact themselves. He says:

> CSIRO believes it is essential that Australians feel comfortable with their understanding of this technology in order to decide about the role it will play in their lives. [We believe] it gives us the potential to improve our health, create a safer and more secure food supply, generate greater prosperity and attain a more sustainable environment.

Of these good intentions there is little doubt in my mind (though I am puzzled by this thing called 'sustainable environment'). Dr Raven and Dr Peacock are undoubtedly driven by concerns as conservationists as well as by the enormous, beguiling attractions of this immensely powerful science. But to achieve their aims they have to do business with wealthy corporations whose public dealings have been less than impressive. The disquiet is more than 'anti-science'. It is a natural response to a world appearing to change faster than it should. It can be, at worst, a bloody-minded resistance along similar lines to Pauline Hanson's

One Nation reluctance to accept change. Or, it could be, at best, tempered by the kind of exercise mediated by the Australian Museum and Consumer Council's Consensus Conference. More on that later.

Of monsters and hubris

Sir Gustav Nossal and others have remarked that nuclear power and genetics seem different from other branches of science in that they foment more, sometimes irrational, anxieties. Both involve invisible, minuscule entities which can disfigure, destroy, make monsters. Promethean forces.

But there are also other parallels. In the beginning, nuclear power was also to be part of our salvation. It would provide electricity 'too cheap to meter'. It would save the miners (like my father) from the deep pits where they dug coal and breathed dust that gave them black lung. Was it not Lenin who said that socialism is 'communism plus electrification'? Was it not George Orwell who wrote: 'In order that Hitler may march the goose-step, that the Pope may denounce Bolshevism, that the cricket crowds may assemble at Lord's, that the Nancy poets may scratch one another's backs, coal has got to be forthcoming'?

Like nuclear power, genetics has been sold with bombast. The hype, according to Ziauddin Sardar, writing in the *New Statesman* (where Orwell too penned his thoughts)

comes a close second to the millennium ... Scientists need to control their tendencies to tell the public, their employers and themselves that they are about to solve all the problems of humanity. Gene technology does not need to be oversold. Rather we need to discuss openly and truthfully its benefits and hazards ... At worst, it may turn out to be the hubris of scientific endeavour that finally yields its nemesis.

Oddly enough this 'hubris' comes, in the main, from the new bind in which science finds itself, the new penury. As state funding for research has diminished so scientists are required to spruik for attention, to sell the advantages of their lines. Otherwise they will not attract funding. If they have managed to secure some corporate dollars then either they, or the ubiquitous, febrile company

PR department, will maintain a stream of unrelenting press releases praising the R & D to the skies. Otherwise the research program may not be renewed. Double bind. Gone is the reticence of yesteryear. Gone is the reluctance to report their findings publicly of even the most accomplished research scientists, who would tell me in the early seventies that public comment on high achievement was premature. Always premature.

But there are undoubtedly Promethean fears associated with genetics. Prometheus, it may be recalled, not only built the first man from clay but later defied a disgruntled Zeus and supplied his creation with fire. The flames can be used to cook dinner and give warmth. They can also destroy. If you think of the first, it is of welcome home comforts; think of the second, and dream of holocaust. The public may well see only the hazards and not understand the cosy benefits. This is shown by Ian Wilmut's surprise at the reaction to Dolly the sheep. The review of the book *The Second Creation, The Age of Biological Control, by the Scientists Who Cloned Dolly* (Ian Wilmut and others) in the journal *Nature* comments:

> Dolly is living proof that the genetic cloning of adult human beings is no longer in the realms of the impossible.

And then:

> Even more ironically, Dolly is most significant by far to the billions of people who least understand her. Around the globe, her image was splashed across television screens interspersed with footage from films of *Boys From Brazil*, *Frankenstein*, *Brave New World*, and *Multiplicity*. Instantly, Dolly's public fame emerged not from what she really is, but from what she represents. She is a metaphor for the Promethean power that scientists now have to create and control life, and her innocent image drives fear into the hearts of those who think man has wrongly crossed into God's domain.

Such scientific hubris has reached what one hopes is its zenith with Craig Venter's announcement that the geneticists in his humungous company, Celera, are close to the creation of a new species from a set of separate genes. These would be stitched together to give rise to a novel germ. In December 1999, in another valediction to a turbulent century, Venter and his team published in the journal *Science* that they had done virtually everything necessary short of such a remarkable creation. Shortly after, in the first week of the year 2000, Venter boasted that Celera had 'greater than 97 per cent of all human genes' in its database. The journal *Nature* inferred that the purpose of this announcement was, in part, to raise the value of Celera's shares — 'giving the company, at least on paper, a market value of US$6.3 billion.'

The journal wondered whether such an 'ownership' of gene sequences is in the public interest:

> Are the interests of the investors and the public at odds? Celera's playing of the market increasingly threatens such a conflict: the financial expectations now being placed upon the genomics industry will necessitate secrecy and potentially burdensome licensing ...

And then the telling reminder: 'Remember Monsanto?'

While companies and their acolyte research associates continue to combine Promethean hubris with rapacious market manipulation, can one blame public opinion, however poorly informed, for inferring biological Armageddon?

A distraction?

I need to be frank. As a science journalist over the last thirty years I have tracked some of the most horrendous violations of the natural world and of the human spirit. From *Silent Spring* and *The Population Bomb*, to Bhopal, Alamogordo, Hiroshima, Love Canal and Minamata; from Terrania Creek, the Franklin River and Maralinga, to eugenics, lobotomies and drug zombidom. Do GMOs rate anything like as high on such a list of outrages? Frankly, no. Not as science. Perhaps, only perhaps, as environmental threats. In terms of nefarious business practice, just possibly.

But never before have I seen so many leading scientists openly flummoxed by the attacks on their research. Many of them chose their fields because they considered themselves to be green. They wanted 'natural' mechanisms to replace the chemical drenches ruining the land and its wildlife. They had also, like Jim Peacock, done much to foster the public understanding of their science. Why were they now being attacked? As agents of Frankenstein? If the public wanted evidence of safety why were they raiding their experimental plots, rendering 'proof' impossible to attain? Why was the public demanding standards of certainty which science could never supply?

Apart from hype (mentioned above) it did not help that the GM foods seemed to have been launched onto the market by stealth. Dr Ariane van Buren, environmental director of the Interfaith Center on Corporate Responsibility in America says, 'Across the US, people unknowingly consume genetically engineered food, even though its safety has not been proved. Already 40 per cent of corn and 60 per cent of soybeans planted in the USA are genetically engineered'. Dr van Buren lists the companies involved in GM products as: 'Monsanto, DuPont, Dow, American Home Products, Novartis, Schering, Hoechst, Rhone-Poulenc, Albertsons, Archer Daniels Midland, Bestfoods, Coca-Cola, General Mills, Heinz, Kellogg, Kroger, PepsiCo, Philip Morris (Kraft Foods), Proctor & Gamble, Quaker Oats, Safeway, Sara Lee and Tricon Global Restaurants (Pizza Hut, KFC, Taco Bell)'. Near Ubiquity.

Stealth, perhaps. But, again, what about the *science*? In February 2000, the newspaper *USA Today*, in a defence of GM foods, headlined an editorial, 'Don't Misrepresent Biotechnology'. They averred that 'critics of biotechnology are inundating 18 major corporations with proposals to stop the development, marketing or selling of bio-engineered foods' — this quashing any suggestion that the GM controversy is restricted to Europe, particularly Britain. They note that the European Molecular Biology Organisation, 'made up of 1000 top scientists from 24 nations' had put out a statement saying that 'if it is dangerous to eat "foreign DNA" or protein, we've always lived dangerously, since everything we eat contains foreign DNA and protein'.

Do people, even Americans, really believe that they are at risk of infection by feral genes from tomato paste or soybeans? How can this risk (if risk there be) compare to the daily hazard of breathing or kissing? We eat DNA every day. We simply digest it. As for alien genes, every time we are bitten by a bug or a mosquito we become (usually briefly) infected by new genes.

USA Today, however, gave three reasons for concern (based on the European Molecular Biology Organisation's statement) about the 'fight' over biotechnology and its products:

1 'It leaves in place conventional farming methods that now damage the environment, through erosion, waste of water and pesticides;

2 'It delays the development of a genetically modified rice that world health experts hope can prevent 500 000 cases of blindness and 2 million children's deaths each year caused by vitamin A deficiency;

3 'It drives away investment in biotech crops and medicines.'

To this list Sir Robert May, the newly elected president of the Royal Society adds this caution:

> Today's intensive agriculture is not sustainable in the long run. In developed countries, we typically spend ten calories of fossil fuel energy, in various ways, to put a calorie of food on the table; a century ago this ratio was one to one, and in hunter–gatherer times was 0.1 to 1. At least half the atoms of phosphorus and of nitrogen incorporated into new plant material around the world today comes from fertilisers, rather than by natural processes. But we cannot turn back the clock, because we could not feed today's global population, much less tomorrow's, with yesterday's agriculture.

Sir Robert warns that this challenge, a priority with scientists, may be a 'two-edged sword' for agribusiness. Their desire to make profits could subsume any wider interests, public or otherwise. He was aware, one assumes, of Bill Clinton's recently revealed pressure on Tony Blair to help reduce the European market's resistance to GM produce.

Is the GM controversy, then, causing us to ignore other, more pressing concerns? Of this I have no doubt. Who is to blame? First in line are the media. There is a strong element of fashion in all news, not least environmental news. Dolly and Frankenfish are *new*. Population, biodiversity, global warming, ozone holes, and so on are *old*. It also helps that there is now extreme uncertainty about politics; the old enemies have faded and there are no obvious good guys. It is still easy to be fearful of Mammon, often with good reason. If, therefore, Dr Frankenstein is working for *Fountainhead Inc.*, so much the better. Or worse.

John Gray, Professor of European Thought at the London School of Economics suggests that, paradoxically, it is the traditional conservative groups who are most responsible for the universal public disquiet. In the old days, when asked what kind of 'society' they favoured Conservative politicians would say 'not communist' or 'not socialist'. It was response by elimination, not prescription. They had no conception of the growth of the general community either, says Gray, because they had not thought it through or because they felt the old-fashioned emblems of family, village green and church sufficed. Now, however, we have the greatest force for the destruction of good old bourgeois values ever unleashed: powerful new technologies combined synergistically with an almost unfettered marketplace.

'Conservative parties seek to promote free markets, while at the same time defending "traditional values". It is hard to think of a more quixotic enterprise. Free markets are the most potent solvent of tradition at work in the world today. As they continuously revolutionise production, they throw all social relationships into flux ... Nothing in the development of capitalism ensures that it is compatible with an intact bourgeois civilisation. The combination of slash-and-burn Anglo-American capitalism with unprecedented rates of technical innovation is particularly inimical to bourgeois life. When a stream of new technologies floods through deregulated markets, the result is not social — or economic — equilibrium. It is to throw the social divisions of labour into flux.'

Little surprise that alienation is well-nigh universal. This alienation is shown by the volatility of the electorate, especially in Australia, and the dumping of several insouciant state premiers who had seemed installed for life. It is shown by a pro-monarchy movement during the campaign for an Australian republic in 1999, which was able, through several government ministers, to convince voters that politicians shouldn't be trusted. And it is shown in American politics by presidential hopefuls from Ronald Reagan to John McCain campaigning that power should be reduced in Washington (where they intended to assume it).

Now to try to answer the questions about genetic science I posed at the beginning:

1   Is the GMO debate being conducted in a manner likely to enlighten?

My answer is 'No!' It is being done better in Australia than in Britain but it is presented in a way that bundles the whole of genetics into a basket. Critics of GM are demanding a rejection of a whole field of activity. This would be equivalent to asking for a ban on chemistry. It is anti-intellectual and against the public interest.

Another way to put the question would be: *Can* the GMO debate be conducted in an enlightening manner? To this question the answer is a resounding 'Yes'. The proof is the success of the Consumer Council's Concensus Conference of fourteen citizens who were able to sit for three days of expert presentation (over which *they* had control) and then to offer a list of supremely sensible recommendations. They had understood the technicalities and quizzed the experts but had not been intimidated by them. The essence of their conclusions was to ask for transparency, consultation and regulation in the public interest.

2   Is the science best separated from the finance?

Answer: yes, but not permanently. Science is never value free; it is hardly ever dollar free. But the hazards of any application of science depend on the openness of the democratic process and the control the community has over applications. The behemoths of the GM trade have only themselves to blame for the contempt and fear with which they are regarded in Australia, Britain and, less so, in the United States.

Each case of genetic modification should be considered on its merits — just as each therapy in medicine is uniquely scrutinised. Every aspect will have risks. Most will be small.

3   Is there a qualitative difference between GM and other environmental issues?

Answer: 'Yes'. The ease with which newspapers and anti-GM lobbies can invoke scary monsters is legion. It means that vitally important (literally!) work on, say, stem cells, could be jeopardised because of the universal smear of Promethean tinkering.

Nuclear research has a similar difficulty — invisible forces of unfathomable power can suddenly rebel and wreak vast destruction. But the nuclear debate was led, from the beginning, by eminent scientists — Einstein, Bethe, Oliphant, Oppenheimer, Pauling, Sagan, Ehrlich. I can think of very few geneticists (David Suzuki?) who wish to lead the charge against GM. Most of the radical scientists I talk to — Stephen Jay Gould, Ehrlich, Lewontin, Steve Jones — are appalled by the conduct of the public debate.

4   Could GM distract from major environmental concerns?
Again: 'Yes!' This is the 'Century of the Environment'. It should be driven by rationality, excellent scientific advice and a proper sense of priorities. Not by fashion. There is a strong element of bourgeois individualism in the search for the 'natural', non-threatening tomato. There is the high church pantheism of Prince Charles in his Reith lectures. There is also the temptation to ride on the ever-simmering Hansonist disaffection with authority and to exploit populist credulity. The case for conservation is utterly overwhelming. It would be tragic to have any part of it devalued by GM hysteria.

***

In volume three of the 1818 edition of *Frankenstein*, the monster meets his maker (and father!) and pleads for understanding and the right to be treated as a normal human creature:

'Shall each man,' cried he, 'find a wife for his bosom, and each beast have his mate, and I be alone? I had feelings of affection, and they were requited by detestation and scorn. Man, you may hate; but beware! Your hours will pass in dread and misery, and soon the bolt will fall which must ravish from you your happiness for ever ...'

And finally, ominously: 'It is well. I go; but remember, I shall be with you on your wedding night.'

The scientist, Dr Frankenstein, having carefully and with such hopeful ambition created a miracle, and the closest thing to a son, is revolted by his work and now risks the ruin of his new relationship and prospects for a thriving family. If only he had been more fatherly and understanding towards his handiwork. Mary Shelley's message (and there are many tied up in the extraordinary novel) is very much one of enlightenment over prejudice.

Hers was a society in upheaval and transition. Ours is the same in many ways but we have the benefit of far more knowledge. We too are desperately uncertain of our fundamentals, not least among them the role of powerful corporations in our hope of progress.

When I interviewed Chris Leaver, Professor of Plant Sciences at Oxford University, about his newly created, genetically modified potato capable of producing twice the normal amount of starch and an excellent prospect for a hungry world, his exuberance was blighted by the latest GM upheaval. Arpad Pusztai's paper was about to appear in the *Lancet* and Professor Leaver was speaking for the Royal Society on GM affairs. At the end of our discussion, trying to account for the flak, he looked despondent and remarked: 'We all live in a capitalist society. Unfortunately!'

2 | POPULARISE OR PERISH?

A new opinion poll has found that more Australians are interested in news about science, technology, health and the environment than in sport, politics crime or employment. The remarkable thing is that this is the fifth poll to reach this conclusion in eight years ...

The quotation is from an article, 'Inside the Australian Media', in the *Walkley*, the journal of the Media Alliance (the journo's union). It is by Julian Cribb, once the leading science correspondent for *The Australian* and founding President of Australian Science Communicators, now in charge of media relations for CSIRO.

He continues: '... yet Australian news coverage of sport, for example, continues to overshadow reporting of science by a ratio of 100:1. Why the mismatch?'

There have been more than five such polls. I have seen over a dozen in the past twenty years, from the BBC, the *New York Times,* the *Sydney Morning Herald* and others. One must assume that the polling manages to exclude vainglorious self-descriptions that allow those questioned to pretend brainy preoccupations over self-indulgent ones. So, assuming that the polls are right and, as Cribb avers, editors appear to ignore the prescription, one might infer that newspapers and other media are not interested in circulation or ratings. As this is false, one must search for other explanations.

It helps to ask personal questions. Despite my own need to keep up with what is written and broadcast about science I find the prospect of sitting out another furry animal and space rocket documentary or having to grind through another dense article about neutrinos or geriatric piles as enticing as flossing my uncle's teeth. However worthy the subject it needs to be inviting. This is especially so if the topic is over-burdened with necessary information. There is nothing intrinsically compelling about any scientific subject — even stories about saving your own life can be made tedious. Much newspaper coverage of science is of three types: *Mondo Carne* (what a weird world it is!); *Frankenstein* (here comes disaster); or *Arcania* (here's something you've never heard of before — and now it's proved wrong!).

Mondo Carne is the daily standby of newspapers and television. Here are a few examples made up from my own material:

- 'GM plants will grow their own greenhouses on Mars' — this was proposed by Freeman Dyson of Princeton University; 'Read your e-mail from your sunglasses on the beach' — *Time* magazine forecast; 'Cama produced by mating llama with a camel' — announced by Professor Roger Short from Melbourne; 'Dolphins can detect trickery' — from the *London Times*; 'First publicly admitted human clone by 2004' — Arthur C. Clarke forecasts (he also expects computers to be connected directly to brains by 2025; 'Other universes predicted' — Sir Martin Rees, Astronomer Royal.

Frankenstein:
- 'Bodyless monkeys kept alive for several minutes' — Professor White of America still at it with assistance, no doubt, from Igor; 'Scientists discover gene for ... aggression–depression–repression–regression–death–infidelity–pole-vaulting–macrame–map reading–sloth.' — delete as applicable; 'Frankenscience: Surgeons have given a man hands from a dead body. Will leg, face and body transplants be next?' — *The Australian* (*London Sunday Times*).

Arcania:
- 'CSIRO to get small budget increase'; 'Sir Wenvil Podger dies'; '803 000 year old hand-axes found in S. China'; 'Australian Industry Starving R & D' — headline in *Nature*.

Many of these (not the ones made up) will appear in the programs I present or the articles I write. They are categories of science story which editors recognise. They regularly appear as tail-end-charlie weirdo reports in current affairs programs. They will sometimes turn up as leads if deemed sufficiently shocking.

Executive Producer of leading current affairs show:
    Robyn, they've just announced in America that they
    can create new life from separate genes.
RW: I know. Craig Venter from Celera.

EP: Are you doing anything about it in 'The Science Show'?

RW: I did it eight months ago. With Venter. I fact I had him on the promo saying the very words 'we can make new life forms'.

EP: But it's news now.

RW: So it wasn't news when I did it? Was it olds?

EP: It's going to be in the papers tomorrow, on television tonight with footage from America.

RW: So because it was only on an Australian program nearly a year ago nobody noticed and it didn't count?

EP: That's how it works. D'you have any suggestions how we can cover the story? We go to air in twenty minutes.

RW: I have an interview with the bloke who was with Watson and Crick and ran the Cambridge Molecular Biology Department saying it won't work because you need genes which are compatible and synergistic, not a glorified genetic Lego set.

EP: Can we run it?

RW: It's ready now!

And so it was, in a glorious re-enactment of the opening of the movie *Broadcast News*, we had an interview I had recorded nearly a year before breathlessly introduced as the lead item of that day's prestige news show. It had been part of a lengthy, absorbing (I hope) discussion of genetic engineering and the future. Then it had been transformed by truncation and setting into a story combining all three prime editorial qualities *Mondo Carne* (they can knit a genome), *Frankenstein* (they can create life), *Arcania* (genes need to evolve together if their sum will work as an organism).

Any scientific story can be expanded, given proper context and so be allowed to grow from its diminished category as a news item. It is likely that the incessant parade of overblown *Mondo Carne/Frankenstein/Arcania* stories leads to confusion rather than edification among the general public and consequently to a hostility to science itself. So much is promised — so little seems to be delivered. So much hazard — the world is becoming an increasingly hostile and even dangerous place.

This is not the impression given when science stories are given space to breathe. The seriousness is not diluted, simply given room to make sense. This is best done in the form of narrative with good writing and tight editing, making the reader, listener or viewer not do *less* work but get sufficient reward for attention given.

So this might be one answer to the conundrum: why doesn't the public respond to science news in the ways that polls predict they should? It's because the science offered is not the sort imagined in the minds of those polled. It has been devalued. One way to solve this problem will be dealt with later — it is the *Multiplier Effect*.

Bad talent?

Science communication is replete with myths. One favourite is that scientists, as a body, are poor speakers and writers. This is nonsense. Has anyone done a controlled study? Have chemists been compared to actuaries, geologists to plumbers or physicists to undertakers? Who are the excellent communicators in our well-spun society? Politicians? Lawyers? Grocers? How have some of us managed to run weekly science programs for many decades (high rating ones at that) if the prime ingredients are all jargon-crunching mumblers? How is it that the Goulds, Dawkins, Goodalls, Leakeys, Suzukis, Davies, Margulises, Ravens, Diamonds and Greenfields can (if they ask) command US$10 000 upwards for a public lecture? Why are the science books by these authors and many others some of the best sellers of the last twenty years, sometimes yielding US$2 million in advances?

Many scientists are superb communicators. We (the Australian Museum) invented a Eureka Prize for the promotion of science eleven years ago. There has never been a shortage of candidates from within Australia. Quite the reverse. All winners have been from the first rank of research, not greybeards choosing the limelight instead of the bench to occupy their twilight years. Science Now!, which took the place of the annual ANZAAS congress, invented *Fresh Science*, a competitive selection of young Australians who present their work during National Science Week. One hundred apply, sixteen are chosen. They are given tips on presentation by members of the ASC (Australian Science Communicators) and then let loose on stage and in the media. They are uniformly brilliant performers.

Yes, the training of scientists militates against fluent use of the vernacular and traditional caution is an additional brake. But protocols are now being relaxed and publishing is being done on the Internet. It is also stated policy to encourage the public awareness of science (see below) and in such a setting we are seeing scores of Australian scientists venturing boldly where once no boffin dared. To say scientists are bad communicators is simply out of date. I doubt it was ever up to much. Don't Australians remember Crosbie Morrison, Douglas Mawson, Macfarlane Burnet, Harry Messel, Peter Mason and all the other local stalwarts?

There is, however, a penalty extracted from those too willing to accept high profile. Many academics are penalised by the system for giving their time to public demands. Dr Rob Morrison from the University of Adelaide has argued eloquently against the lack of formal merit points awarded to university scientists (which count towards promotion) prepared to speak out. Ian Anderson (Australian editor of *New Scientist*) has written that 'over and over again, scientists are chastised — not by the public or the media, but by their colleagues, who should be offering support, not criticism'.

A spectacular example of a scientist suffering for the cause of popularisation was Carl Sagan. Jared Diamond, in an article in *Discover* magazine, tells how Sagan had the grace to rise above insult, even when the insult came from the American Academy itself:

He described how he too, had taken flak from other scientists, but — he paused, as if to choose his words carefully — the disadvantages to him had not been serious. As he uttered these words, I sensed my fellow academy members holding their breath, waiting to hear whether Sagan would mention a stinging insult he had suffered at the hands of the academy members themselves. In fact, he passed tactfully over the scandal that had unfolded a few years earlier, when he had become one of the few people in the academy's long history to have been provisionally elected to membership but then individually rejected in a special vote.

Sagan knew that day that he was probably dying. He refused to give himself a wild card and indulge any desire for retribution.

Sagan could be arrogant and inclined to hubris; this is well documented in Keay Davidson's recent biography. But Sagan's rejection by the academy was as much to do with the unacceptability of profile as it was to do with the astronomer's personality. Diamond himself told me he had tempered his own inclinations to popular writing and speaking until his academic credentials were soundly established. About Sagan he concludes:

> Of course there will never be another Carl Sagan, and his loss seems doubly painful because we so badly need scientists with his skill. Just one would not be enough: we need thousands. But we are never going to get them — not until scientists and their organisations drastically change their behaviour.

This was endorsed by Bill Clinton in a statement (Science in the National Interest) saying:

> The lifelong responsibilities of citizenship increasingly rely on scientific and technological literacy for informed choices. Our scientific community must contribute more strongly to broad public understanding and appreciation of science. Our education system must provide the necessary intellectual tools at twenty-first-century standards ... Federal agencies will encourage research scientists to use their research experiences in support of public understanding and appreciation of science.

Australia too is blowing the science awareness trumpet at the highest levels of government, as we shall see.

Are scientists hard done by?

A survey commissioned by the Federal Government when ex-science teacher, Ross Free, was minister concluded that young people felt scientists were 'nerds and losers'. Several of us were called to the cabinet room to offer wisdom to the Prime Minister's Science Council about ways to improve the poor boffins' image. (The Prime Minister, Bob Hawke, could not attend that day because he was about to be toppled). No one appeared to be willing to counter the accusation (other than me), only to offer remedies. There is a clear danger that scientists talk themselves into nerdiness and that the political leadership is complicit. While it is true that there are not many scientists who resemble Indiana Jones, there are many whose working lives are far from ordinary: Jane Goodall in forests with apes, Douglas Mawson fighting for survival in the blizzard, Sally Ride in space, Jacques Cousteau in the depths, Mike Archer in fossil cornucopia, Tim Flannery in Papua New Guinean jungles, Sylvia Earle swimming with whales ... the list goes on.

The PM's Council concluded that the image of scientists as weirdo buffoon or autistic fiddler with flex was engendered by the media. Why were there not popular dramas — soap operas — showing how science really is? they wanted to know. This formula was attacked by Michael Crichton (medic, anthropologist and novelist) in a paper given to the AAAS in Philadelphia in 1999:

> Scientists often complain to me that the media misunderstands their work. But I would suggest that in fact, the reality is just the opposite, and that it is science which misunderstands the media.

Then the man who wrote *Jurassic Park*, *ER*, *Timeline*, wondered whether scientists had given much thought to they way others are depicted.

> The implication is that scientists are singled out for negative portrayals, and that the public is therefore deceived in some way we should worry about. I say that's nonsense.
> Let's be clear: all professions look bad in the movies. And there's a good reason for this. Movies don't portray career paths, they conscript interesting lifestyles to serve a plot. So lawyers are all unscrupulous and doctors are all uncaring. Psychiatrists are all crazy, and politicians are all corrupt. All cops are psychopaths, and all businessmen are crooks. Even movie makers come off badly: directors are megalomaniacs, actors are spoiled brats. Since all occupations are portrayed negatively, why expect scientists to be treated differently?

Crichton has a point. The problem is that, having extracted a perceived negative image from popular culture, scientists themselves have been responsible for magnifying it. Had they left themselves quietly in the movies along with all the other caricatured

professions they may have transcended the smear. But even when
they are the heroes they are not seen as such. There is a myth that
the Richard Attenborough character in *Jurassic Park* is a nutty,
bungling scientist along the lines of Dr Frankenstein. But he isn't:
he's a businessman. Crichton explains:

> The other two people in the picture are scientists and they have nothing to do
> with the bungling ... In passing, I'd remind you *Jurassic Park* does have a sci-
> entist as its hero. Alan Grant. He saves the kids, he saves the day, rights the
> wrongs, and looks dashing. Beside him is another hero, Ellie Sattler, a
> botanist. So in a movie where nearly everyone has a doctorate, why talk
> about wanting to be heroes not villains? The scientists already are heroes.

There is a certain duplicity in Crichton's argument, however.
He says that scientists are fitted, like everyone else, into
Hollywood's mill: they don't get special negative treatment. But
the point is that Hollywood need not have only one mill, telling a
story in one formulaic way. Already there is a recoil from the high-
dollar monopoly that the American film industry has forced onto
the world audience, using outer reaches such as Canada, Australia
and even Britain only as service depots for special effects, settings
or ready-made acting talent. The cheaper techniques now available and multiplicity of television channels should enable many
more film-makers to provide drama alternatives in the future.
Science does have a problem of speed — which will be addressed
later. But Crichton avers that most of the complaints about the
distortion of science are based on myths:

> Mass media isn't mass. Mass media isn't respected. Mass media isn't influ-
> ential. The media has lost its power. All the more reason for science to stop
> worrying about how it is portrayed.

If Michael Crichton's argument has any value at all, it is to
convince everyone that the time has come to stop presenting scientists as whingers. The culture of complaint is unproductive for
nearly everyone as public policy. It is better to build alliances than
create adversaries. Modern politicians are skilled at playing sides
off against each other. It may not be fair but it is a reality. Ask any

unprivileged group, the poor, the downtrodden, the powerless, how much attention they gain through their incapacity. It may happen, briefly. But it doesn't last. Now, more than ever, the 'Matthew Effect' applies: 'Unto them that hath, it shall be given.'

The suppression of science

I have worked in the ABC Science Unit since March 1972. I have presented the Radio National program, 'The Science Show', since it first went to air in 1975. This may appear to be obsessive–compulsive behaviour, given the mixed attractions of the national broadcaster which can sometimes resemble a cross between a war zone and a garage sale. On the other hand, if you accept that science broadcasting is a legitimate career option and that you don't just grow out of it when something more mature-minded turns up (such as management) then the questions arises: where do you go in Australia if the ABC is no longer an option?

This question is usually met by silence. 'Beyond 2000' turns up now and then on commercial television, sometimes with stories which have lingered on the shelf for two years and which are put on to fill gaps. Despite the doubling of the ratings and the surprise it engenders, the program does not resurrect in prime time. 'Sixty Minutes' does science now and then, usually through the enthusiasms of the admirable Jeff McMullen, and there is a health program on Channel Nine, but it tends to concentrate on the aches-pains-and-therapies end of the spectrum, not research. 'Burke's Backyard' often gives a good fist to biology and medicine. Other channels have vets and pets, deep nature and panel chats. If you define science journalism as the attempt to bring a broad range of scientific topics, however challenging, to a public deserving to know more than gloss, then it's either the ABC or nowhere.

In 1996, I wrote a book, *Normal Service Won't Be Resumed*, that tried to explain the difference between 'specialist broadcasting' (taking pains to investigate topics and issues, then applying high production values to the resulting programs) and 'flow programs' or 'chat shows'. The latter dominate the networks and are virtually the only kind occupying commercial airwaves on radio. The reason is obvious: they are cheaper. By many orders of magnitude. To present a chat show you develop a compelling, often obnoxious on-air personality, open up the phone lines and leave the door open for the succession of guests wanting to flog a product (sometimes themselves).

I have the honour to be president of the Australian Science Communicators (ASC), the professional organisation for science writers and broadcasting. The ratio of journalists to public relations officers in our membership is 1:10. For every ten ASC members sending out press releases there is but one to receive them. I clear the fax machine much of the time in the ABC Science Unit and it is with exasperation that I note we receive several half-kilo blocks of unsolicited and unwanted notices every day. At a recent conference of public service public relations personnel we were told that 97 per cent of releases end up being used somewhere (publications are always looking for ways to fill space cheaply), sometimes even two years after they were sent out! Many institutions are competing for attention in a tough environment and no editor looking for cheap fill need look long. The idea of actually doing some journalism to follow up the release is becoming almost quaint. This is why someone once said, 'there aren't science journalists, only stenographers'.

Even within the ABC it has become a struggle to mount a comprehensive science coverage. There are vehicles such as 'Quantum' in which a few stories can be placed. Beyond that, the trend is either to appease the intense desire to attract a younger audience through the cheap banter of panel shows or to face the hegemony of the international market with the production of special series. The ABC cannot afford to invest in too many programs unless there is a good chance they will be picked up by Europe and America. Each zone imposes severe demands: no Australian presenters, no pesky boffin talking-heads, no awkward topics. This triage is hard to face. It is easier to offer the bland glossy fare one knows the international market prefers. Once, when a man from *National Geographic* was preparing for his first tour of Australia and asking me for tips, I sent him a list of front-line research ideas and he responded, 'Nah, we don't go for that jazz. We want typical Aussie stuff on koala bears, kangaroos and Barrier Reefs. Can't you send me some of that, Robyn?'

In America the PBS (Public Broadcasting Service) network is limited by sponsorship and presenting shows that will attract subscribers. In 1999 the *Washington Post* accused PBS of skewing its programming away from solid science towards New Age medicine and even shows offering tips on immortality. This is an extract of a report, actually by PBS exposing its own recalcitrance:

> Who's the biggest star on PBS? Jim Lehrer? Jim Glassman? Well, sorry Jim. If pledge drive contributions are any measure, then no one can hold a candle to this man. His name is Gary Null. His self-help special, 'How To Live Forever', is a pledge drive bonanza ... In that program, Gary Null, who is not himself a medical doctor, says we can reverse the aging process by 'detoxifying your negative emotions'. Not exactly peer-reviewed science, and Null is not the only pledge week star offering health tips. In his pledge week special on alternative medicine, Dr Andrew Weil laments the decreasing role of the shaman, and says that hypnotherapy and visualisation therapy are valued treatments for rheumatoid arthritis.
>
> As effective as Null and Weil have been as fundraisers, their role as medical advisors has raised serious questions. Critics say PBS stations have made a Faustian bargain, elevating fundraising above their educational mission. They say it is irresponsible for a network that airs solid science programming such as 'Nova', to air pseudo-science programming on touch therapy and bio-magnetic healing.

Public broadcasters are facing a continual succession of funding cuts. Science broadcasting is expensive: research is required and difficult questions must be asked and pains taken to make programs substantial instead of superficial. If America cannot afford science on radio and television beyond the gloss that sells easily, then what hope is there for Australia?

Paul Farhi of the *Washington Post* comments:

> Even the head of PBS, Irvin Duggin, in an e-mail that I find extraordinary, says 'we raised the bar by putting on programs like 'Nova', we honour science by putting those programs on and then we allow quacks and charlatans,' those are his words, 'on our air. What are we doing to ourselves?' ... That's the real point. They've got to keep the doors open, they've got to pay the light bill ... There have got to be other sources of funding. They don't have much choice.

Just as there are agencies keen to get certain messages to air and into the press, so there are others trying to keep them out. It is impossible to tell how many important science stories are being suppressed because, as I have indicated, there is less opportunity for journalists to spend the time required to investigate a complex issue. It took Norman Swan five years to bring the report of Dr William McBride's scientific fraud to the public via 'The Science Show'. Some disturbing examples of suppression do, however, surface now and then.

The name 'Monsanto' tends to turn up quite often. On one occasion I was called by the famed herpetologist Professor Mike Tyler who felt his linking of the demise of frogs all over the world with the herbicide *Roundup* was being attacked. Dr Tyler had received a series of threatening phone calls. He could not identify the callers but told me that they invited him to desist from his public statements suggesting that the surfactant in *Roundup* was responsible for killing frogs. We recorded an interview detailing his concerns and I put it in a drawer — just in case. When Tyler was convinced his evidence was safe, we broadcast most of the tape. Monsanto did not contact me or give any direct indication they wanted to interfere with the journalistic process.

This was not the case, apparently, with Jane Akre and Steve Wilson's report on the alleged effects of Monsanto's synthetic hormone r-BGH found in the Florida milk supply. They told Mick O'Regan on Radio National's 'The Media Report' that they had not only been stopped from broadcasting the story on Fox TV, they were sacked for their pains.

JA: We found that on seven out of seven farms that we visited, the farmers were using it, that the University of Florida was a big promoter of this drug and they were at the same time taking more than $1 million a year in gifts from Monsanto. We found that there was a growing line of evidence that said that a by-product of the use of this hormone is linked to a proliferation of tumours. And we found some very good experts, some highly credentialled experts who went on camera to talk about that.

SW: Well, at first, when we wanted to do the story and then got this tremendous pressure from Monsanto, the first thing they did was they decided they just wanted to sweep it under the rug. And they said to us, 'Look, if we do that and we kill your story, will you ever tell anyone?' And I looked at the station manager and said 'Only if they ask ... And on two separate occasions they offered us US$200 000 essentially to shut up and go away; again barring us from not only talking about this dairy hormone, and for that reason we decided we wouldn't take the money.

M O'R: So what's this revealed to you about the way in which news is managed in North America and possibly in other countries like Australia?

JA: I think to stay in the business you have to put up with it and shut up, and that's the way you keep your money flowing in and your mortgage, and your car payments, and that's what you have to do; you have to play the game to get along. And heaven help you if you stand up for good journalism, because you're going to find who your real friends are, and there are very few of them.

SW: And you know, really more disturbing than that, it says that news has just fundamentally changed. It's become a business, just like making lightbulbs. Decisions which used to be made based on the public's right to know and the public's need to know things to carry on their lives and to make good decisions about what they put into their bodies, things as fundamental as that, news decisions are now made based on what it's going to cost, not based on the old way which was what people needed to know and have a right to know.

This kind of reporting is a long way from *Mondo Carne/Frankenstein/Arcania* though it does have elements of all three. The philosophy of 'The Science Show' has always been to adapt the length and style of the coverage to the essence of the material being dealt with. This means that the program is unpredictable and demanding. Where else can this be achieved other than in a public broadcaster such as the ABC?

All three categories (*Mondo Carne/Frankenstein/Arcania*) are of science in a postmodern marketplace. Like most media items they are offered as entertainment, shock or scandal and so resemble other objects offered for speedy consumption rather than for reflection. The critical difference between the 'science' the public claims it prefers and the science I or other science journalists (and even scientists themselves) provide is therefore the difference between what is titillating and distracting versus what is interesting but takes some effort to consider. It is the difference between a factoid and an idea.

Providers respond to the postmodernist market imperative by supplying panel shows, sound bites, bright paragraphs on websites and small books of trivial pursuit and jokey science. Extended programs such as those I and relatively few other survivors now broadcast have audiences ten times smaller than when they began. It is not that these programs are rejected by the newly discerning, better-served Australians. The shows are unknown. They are swamped by the tidal wave of ephemera in the new age of promiscuous communication. Outside Australia they are almost nonexistent.

Budgets are small, outlets proliferate, therefore content must be turned over quickly. Only sure-fire subjects such as dinofests ('Walking With Dinosaurs') or space extravaganzas ('The Planets') justify taking pains in the new market. Ninety-eight per cent of the science I and magazines such as *New Scientist* offer is not scandalous, shocking or amusing — only interesting. And it is our duty to display this range of subject matter (remember media of record?) not just the juicy bits. The crucial difference between the public which says it wants more science and the actual audience/readership editors know is one of constancy. It takes practice to listen to 'The Science Show' or to read *New Scientist*. Today's younger public will: 1. not know where to find either; and 2. not stay the course to test the product, except in rare circumstances. Everything about their training and the other strident demands on their attention takes them elsewhere.

The pity is that science, like other aspects of culture, cannot be appreciated on the run. It is the same as the difference between art and decoration, opinion polls and democracy. All sensible promoters of the public awareness of science want it to be fun and even games. But not *only* fun and games.

I have no doubt that the public, here and abroad, is perfectly capable of appreciating both kinds of science, the amusing as well as the ideas-in-depth. But the public is not given the chance. A free-for-all on the net is not the answer.

The future

The public awareness of science is said to be a high priority in Australia. Gareth Evans, with the implicit backing of the Department of Foreign Affairs, said so at the UNESCO World Science Conference in Budapest in July 1999. Dr Robin Batterham said so when appointed as Australia's chief scientist in the same year. The Prime Minister, Mr John Howard said so on 5 December 1999 when speaking at the PM's Science Council at which, once again, there was a session dedicated to the issue.

A group of us, led by Dr Vicki Sara, Head of the Australian Research Council, spoke in the cabinet room about initiatives we recommended should be taken by the Federal Government. John Howard was present throughout which seemed to confirm his oft-stated claim that attending the Council (which sits for one day twice a year) is a highlight of his job.

Our presentation included several themes:

- that the public awareness of science is essential to the demo-cratic process;
- that this is especially a priority at a time when science and technology are changing the world at a staggering rate;
- that the approach should not be narrow and aimed only at recruiting science professionals but that science is for every-one in every job;
- that, consequently, it is counter-productive to complain that Australians know more about sports people than scientists and so we should, instead, show how much sport now depends on science to function;
- that 'heroes' from all sections of Australian life should be enlisted to affirm the importance of science in what they do;
- that the prime minister accept the role of a spokesman to promote the importance of Australian science and to do so at every opportunity.

The several ministers present, Minchin (Industry), Vaile (Trade), Kemp (Education), Alston (Communications), Hill (Environment) and the prime minister, as well as members of the Council (various professors, heads of academies and the chief scientist) and, indeed, British Chief Scientist, Sir Robert May, did not quarrel with our line. They actually applauded. Whether this reception will be graced by actual results is a different question. What does seem secure is that the tenor of the approach was right: not to start from a position of victimhood ('they only love Shane Warne not Gus Nossal') and neglect but to offer positives and opportunities.

One disappointment was to be reminded of the limited extent of politicians' horizons. They had not heard of our experiments to bring science to the public in novel ways, despite the fact that they are paying for some of them. The philosophy behind 'Science in the Pub', an initiative of the New South Wales branch of ASC, is to go where people are and listen to what they have to say — a reverse of the ivory tower. Two speakers, chaired by Dr Paul Willis, a palaeontologist now with 'Quantum', speak on the chosen topic and then field questions and comments from those in the bar. It is all recorded, edited, then broadcast on Radio National. This gives a local event a national profile. The transcript can then go on the ABC website, giving it international exposure. This is the *Multiplier Effect* I mentioned earlier: the investment of resources is amortised by repeated exposure in different forms in different media. Members of the public can interact at the event and on the Net. Our assumptions of the subject matter are open to challenge at each point. This is far from sages on stages and admirably suited to a society suspicious of experts, institutions and the authority of VOMITS (Very Old Men In Ties).

The *Multiplier Effect* can also reduce the overall cost of specialised broadcasting. A highly produced report can be distilled into a useful fragment and broadcast on metro media with lighter formats and both can include websites with information about the original, uncut material and books used as bases for the story. The main difference between science and sport, that of speed, can therefore be used to advantage. The attraction of sporting events is that a clear contest is flagged and the result is forthcoming within minutes (horse-racing), hours (football), or, at the most, days (sailing, cricket). Scientific developments are more intangible: the growth and adaptation of ideas takes time, results are never clear-cut and the issues are resolved only many years later. Presented as *Mondo Carne/Frankenstein/Arcania*, science becomes gobbits of disconnected information; presented as multiplied aspects of ideas, science settles into a comprehensible, continuing narrative. You don't need two sides, simple rules and a man with a whistle to make it work. Science, like life, becomes a story you turn to regularly, as a matter of course.

The most important multiplier for Australia could be, in the virtual absence of a unifying body such as ANZAAS, a connection between ASC, the academies, CSIRO, the universities and teaching bodies as well as FASTS (the Federation of Australian Scientific and Technological Societies) as a mechanism to hold the groups together to prevent the endless reinvention of wheels. Another would be the proper integration of science into professional training. Can there be a job left requiring no scientific knowledge? Which lawyers can afford to ignore DNA or definitions of madness? Which farmers need not know about soil science or GMOs? Which cooks can afford to ignore *Salmonella* or cholesterol? Which banker or manager doesn't need to know about computing? Which cricketer or footballer can ignore sports science? Once the subject is integrated, public awareness will tend to look after itself.

Is there a different science for the people?

Einstein said, 'You do not really understand something unless you can explain it to your grandmother'. Richard Feynman added, 'Physics is like sex: sure, it may give some practical results, but that's not why we do it'.

To return to the puzzle with which we began: why is it that so many surveys place science at the top of the public's media wish-list yet editors see otherwise? Professor Chris Bryant from the Australian National University's Centre for the Public Awareness of Science says, in answer:

> The reading public is conditioned by 'need to know'. Only when it has an inter-est will it turn to the science pages. When it is personally engaged, it will read voraciously and be capable of mastering difficult material with ease. It is the task of the science communicator to increase the 'need to know' and nurture it.

That this task is not being tackled, according to Professor Bryant, is shown by the decline in enrolments in science courses at universities by 13 000 in a ten-year period, during which the number of children finishing year twelve has doubled. This can be a tricky argument. Success, as we said to the PM's Science Council, need not be measured by the numbers of those opting for straight science training. There is no clear shortage of science professionals in Australia. There is, in fact, a shortage of jobs for well-qualified scientific personnel. So, the aim is not to produce more Australian scientists but to have a better-informed Australian community. This can be achieved by integrating science throughout various levels of activity, including education, in general. John Dawkins, when Minister of Education in the Hawke Government, liked to say that the top scientific talent is self-selecting and would always form the cream of science faculties at our universities. (This begs the question that most of our universities will have science faculties *at all* in the near future.)

There have been countless experiments, of which 'Science In The Pub' is just one, which have managed to reach the public's 'need to know' button. Add these experiments to the *Multiplier Effect* and one solves the problem of 'dumbing down'. There is always the connection with more knowledge, further enlightenment, with every bright, brief scientific factoid offered. The *Mondo Carne/Frankenstein/Arcania* nexus is broken by there being extra material available alongside. One-minute nuggets on rock stations come linked with extended offerings elsewhere. Science mentions in sports academies come connected to full courses at adjacent faculties.

The obstacle is not necessarily only in political inertia. The scientific profession needs something of a jolt as well. The rise from negativity and despair will help. The seven-foot-tall writer/doctor/anthropologist Michael Crichton put it this way:

> So let's stop the self-flagellation, the ritual abuse and the hot air, and let's follow some new paths. Science is the most exciting and sustained enterprise of discovery in the history of our species. It is the great adventure of our time. We live today in an era of discovery that far outshadows the discoveries of the New World 500 years ago. In a stunningly short period of time, science has extended our knowledge all the way from the behaviour of galaxies to the behaviour of particles in the subatomic world.

Under the circumstances, for scientists to fret over their image seems slightly absurd. This is a great field with great talents and great power. It is time to assume your power, and shoulder your responsibility to get your message to the waiting world. It's nobody's job but yours. And nobody can do it as well as you can.

3 | INVENTING THE FUTURE

Here is an executive summary of this chapter to save time. Science is good. Science costs money. Australia wants to be well off and civilised, therefore it needs first-rate science. Conclusion: Canberra should give more funds to Australian science before it sinks to the level of English cricket, taking the country with it.

Most of us have listened to this argument in various forms for thirty years. It is very wearing to have to trot it out again. Especially as successive federal governments claim to agree with it. On 10 February 2000 the Prime Minister, John Howard, addressed a banquet in Melbourne during the Innovations Summit. He told how he had asked Alan Greenspan to explain the secret behind America's vibrant economy. The head of the Fed did not pause to think. 'Invest in technology,' he replied, thrice. Our prime minister implied he had noted the advice and accepted its wisdom. Where, one wonders, does talk end and action begin? Many of us have wondered this for a long, long time.

Before examining the arguments for maintaining a strong research infrastructure it's worth re-examining the point of it all. What is science for? There are at least five purposes:

1 to explain the natural world;
2 to create wealth;
3 to inform the democratic process;
4 to protect our interests;
5 to tell us who we are.

To tease those out. The first and most obvious function is the motto of the Royal Society of London: to explain the natural world; to answer the question, 'Why is it so?' The initial task is to chart what is there, to list the ingredients of the natural world. This process will suggest in turn, as it did to Darwin after his great voyages of exploration and collection, what relationships they have and how they came to be as they are. Such research is usually seen by outsiders as 'scientists having fun', not realising that it is the essential underpinning of purposes 2–5. The scientists at

CERN (Conseil Européen pour la Recherche Nucléaire, Geneva) were certainly having fun when they devised a means to communicate swiftly over distances — it led to the Internet. Charles Townes and his friends were certainly enjoying themselves when they invented masers and lasers — for which they foresaw no practical use. How could they, back then, have imagined credit cards and CDs? Sir Harry Kroto wanted to recreate a form of carbon found in stars and in the highly enjoyable process discovered Buckyballs, a new branch of carbon chemistry — the ingredients of everything from mini-computers and nanotechnology to aircraft and space-engineering.

But it is this kind of science which is also the basis for culture: the subject of books by Dawkins, Gould, Hawking and Davies; the object of endless fascination for countless ordinary citizens doing science as amateurs; the inspiration for fiction from *The Foundation Series* and *Frankenstein* to *Star Wars* and *The Matrix*; and, of course, never to be forgotten, the basis for the whimsy of satirists such as Woody Allen and Douglas Adams.

The second purpose of science, the creation of wealth, is almost all we hear about from politicians and commentators. In December 1995, Senator Peter Cook (then Minister for Industry and Science) and Paul Keating (the then Prime Minister) gave us the Innovations Statement. In February 2000, Senator Minchin and John Howard, in the identical jobs and even in the identical building (the Convention Centre in Melbourne) gave us the Innovations Summit. There is no doubt that Australian politics is keen to harness the immense power of R & D — but doesn't fancy paying for it. In the words of the head of the Australian Vice-Chancellor's Committee, Professor Ian Chubb, 'Australia fiddles while the world is learning'.

It is quite wrong, however, to define wealth only in terms of material goods that can be sold or exchanged. Dr Clive Hamilton who directs the Australia Institute in Canberra has made a compelling case that the wealth not measured by GDP: GPI — General Progress Indicator or 'the happiness quotient' — is as important for the well-being of a nation's citizenry as anything sold in shops. GPI includes environment, health and leisure, all with substantial scientific components when maintained effectively.

The third purpose of science, informing the democratic process, is becoming more and more patent, as the first two chapters of this volume have attempted to track. Professor John Gray's concerns about the rate of change and the dismantling of bourgeois society, green groups' worries about GMOs and the wholesale transformation of the landscape, legal and privacy questions to do with DNA and information technology, are but some of the urgent issues with which we all must deal. Scientific education is far more than a means to train more boffins. It is the means by which a mature democracy makes informed decisions about its future. Neglect education and you create a credulous society and a suspicious one, as likely to be hoodwinked by shamans as it is to be exploited by ruthless men in stove-pipe hats.

The fourth purpose of science, to protect our interests, is the strongest argument against simply relying on expertise abroad. Australia is different. Our ecology is unique, our population is unusual. All the new medicines, technologies and applications (GMOs!) need to be assessed by our own independent experts before being allowed free rein in Australia. To do otherwise would be imprudent, often illegal. Feral intruders, from stowaway starfish to clandestine germs, need to be guarded against and, if they have managed to invade, combatted as best we can. Stinting on this kind of science would cost us ten times more in the long run.

The fifth purpose of science, to tell us who we are (and who we are not!) could be the most important purpose, but is rarely remembered. Without science we would not know there is only one human race, that women and blacks are not inferior and that the human occupation of Australia has been both extensive and transforming. The last example has had an immense influence on contemporary politics, debunked the prejudice of *terra nullius* and produced both the Mabo and Wik legislation. Australia is now quite different from the land I first encountered in 1964 and unrecognisable to those who grew up here before that. The Australian legend (the one of reefs and roos and rugged terrain, not of chundering nincompoops and swaggering hoons) is essentially a scientific story. From the first moments of the European presence it became pre-eminent and it has hardly faltered since. (Professor Ian Johnston, from the University of Sydney does not include Aboriginal 'science' in this perspective because it lacked a written record but, if it is added, the extent of our scientific history is even greater.)

Science is not the only way to find out who we are. Certainly not, but it is a vital one. It works best in conjunction with history, philosophy and the arts. It is essential to the synergy.

So how should Australians react to these five points? Should they write blank cheques and become cheerleaders for science? Not at all. The alternative to neglect is not implied license. A mature society knows how to fix goals and keep a watchful eye on all its institutions. Science, for all its mystery, is no exception. But what is the evidence that R & D has been allowed to languish?

First of all, there are the numbers. When, in 1999, the Minister for Health, Dr Michael Wooldridge, responded to the Wills Report on the future of medical research in Australia by increasing its allocation by $600 million, the *total* R & D budget still showed an overall *reduction*. Professor Vicki Sara, who chairs the Australian Research Council, told the journal *Nature* that spending on university research here 'has dropped by 13 per cent as a percentage of GDP over the four years of the coalition government'. The headline given to the article was 'Australian industry "starving" R & D'. This was in response to our nineteenth position out of the twenty-four leading economies of the world for business spending on R & D.

The President of the Australian Academy of Science, Professor Brian Anderson (no firebrand he!), commented:

> The need to boost Australia's appallingly low rate of innovation is now urgent. This challenge is greater than tax reform. We need a long-term commitment from all strata of society and all political parties, or as a nation we face a future impoverished by economic and social backwardness. Australia simply cannot have a prosperous and secure future at the current rate of innovation in our economy.

It takes grim circumstances for a genteel body such as the Academy, led by such an otherwise undemonstrative expert in submarine design, to come out with statements like that. The Academy calculated that an addition A$1 billion over three years would need to be added to our science base if we were to match the commitment of countries such as Britain and perform as we should.

Mark Dodgson, Professor of Business Studies at the Australian National University, agrees:

> ... how many times does this have to be said — the government has to increase its investment in universities and technical education. You can't have an innovative, knowledge economy without knowledgeable, creative people.

But what is the evidence that this kind of public investment leads to an increase in wealth? At one stage, in the mid-nineties, there was a strong lobby arguing that innovation was largely the improvement of existing models and techniques — only 6–8 per cent of it required fresh input from basic science. This view is no longer in the ascendant. In 1997 a study prepared by the National Science Foundation in America showed that '73 per cent of the main science papers cited by American industrial patents in two recent years were based on domestic and foreign research financed by government or nonprofit agencies'. Private companies paid for the rest. 'Such publicly financed science,' the study concluded, has turned into a '"fundamental pillar" of industrial advance' (*New York Times*).

This study is regarded as the most 'thorough examination to date of the scientific foundation of American patents' and shows that tax payers' money is at the heart of 'most commercial innovation'.

The occasional lonely sceptic disputes this argument. Dr Terence Kealey, a genial gingery biochemist from Cambridge University and author of *The Economic Laws of Scientific Research* (1996), wants to see Australia jettison all state funding for applied science. 'Trying to explain these simple concepts to the science policy lobby is hard work,' writes Kealey. 'Consider Japan. In a recent article in the *Financial Times*, John Mulvey of the pressure group, Save British Science, praised its investment in science. Only a British science lobbyist would offer today's Japan as an economic cynosure. Japan grew rich under *laisser faire*, *laisser innover*, and its government only started to fund science a few years ago, helping precipitate its economic decline.'

When I quizzed Dr Kealey about this prescription for Australia, given that reductions of state funding remain stubbornly unmatched by industrial investment, he replied that as great exporters of primary produce Australians don't much need science and could probably dispense with it. He was not impressed by my informing him that our agricultural and geological research remain pre-eminent (especially in CSIRO) largely *because* they are needed to reinforce our export industries. Nor by the news that Australia sees its future as more than as a sheep run or mine site.

The fact remains that Australian research, especially at universities, is substantially underfunded. The crisis on campus has been likened by Professor John Niland (himself a professor of business studies) to a 'natural disaster'. Staff are demoralised and students deeply cynical. There is a stampede away from science among young people. They do not see secure careers in science in Australia. 'The Federation of Australian Scientific and Technological Societies (FASTS) claims good scientists are fleeing Australia in large numbers for better opportunities overseas. And lack of support for science in Australia makes it unlikely they will return,' announced the Ian Lowe in a column headed 'Losing Our Minds' in *New Scientist* magazine. He quoted FASTS president, Professor Sue Serjeantson: 'Every Australian scientist has farewelled friends and colleagues off to better jobs overseas. Conditions are better, research funds are more available, job security is better.'

American science is certainly enjoying a bonanza period. In 1998 Bill Clinton announced the biggest single increase in state funding in history for civil science (defence science once had equal funding) taking it to US$38.6 billion. In 2000 it rose to US$46 billion. Government support for information technology research in the United States rocketed by 28 per cent in 1999, adding US$366 million (making one wonder why Mr Gates and his friends don't fund the enterprise themselves). Al Gore lauded the benefits:

> Our economy has never been more driven by science and technology than it is today. Over the past three years, information technology alone has accounted for more than one-third of America's economic growth. More than 7.4 million Americans work in IT today — and those jobs pay, on average, 60 per cent higher than the average job.

Meanwhile, according to academics in Australia, this country lags seriously behind our neighbours (Singapore, for example) in IT fields such as broadband technology. Some see us as lagging by four years.

Investment in R & D in Britain has jumped considerably since 1998 and the same is true (*pace* Terence Kealey) in Asian countries such as Japan *despite* their economic crisis. In March 2000, the government of Ireland announced the injection of 1.9 billion Irish pounds (about A\$4 billion!) over three years. This new money is such a large stimulus that the Vice-Chancellor of the University of Limerick, Professor Roger Downer, says that some observers wonder whether the system can handle it. If translated to Australian circumstances (Ireland has a population of 3.75 million) Canberra would have to stump up \$20 billion!

Young people in Australia are turning away from science courses in ever greater numbers. According to a study commissioned by the Australian Council of Deans of Science, Year 12 enrolments were down 11 500 in 1999 compared to 1989, with the biggest drop occurring between 1992 and 1996, when the numbers fell by nearly 35 000. 'On the best available evidence, there appears to be a marked decline in the number of students undertaking study of the kinds of science that enable participation in the new technologies,' said the deans' report. 'On the only indicators that we have it appears that there is a decline in the study of science in secondary schools, a poor rate of transition from secondary to tertiary study in science, stagnation in a significant section of the enabling sciences at a time of considerable growth in the tertiary sector as a whole.'

In an editorial, *The Australian* quoted one of the consequences of the perceived decline and offered a remedy:

One high-profile scientist has voiced the fear that physics, for example, might soon occupy a similar role to that of Latin: respected for its intellectual demands, retained in a few universities as a status symbol; but no longer part of the educational mainstream — a prospect he viewed with alarm because our technological system is based on physical principles ...

Within universities science, as a high-cost low-growth area, is a natural magnet for the repercussions of the federal government squeeze on higher education teaching and research budgets. It is up to the Government to provide the chequebook leadership to show that is unacceptable. The universities for their part, must find new ways of outreach to help schools put the sizzle and the meat back into science education.

There are several issues implicit in these bleak paragraphs. On brain drains: one must distinguish between the egress of clever minds and opportunities for them to return. It is a good thing to drain brains *temporarily*. In fact it would be depressing to discover that our bright sparks are not tempted overseas. A spell abroad is a near essential part of any good academic's career, which is why we actually need *more* postgraduate travel scholarships. What is important is that there be sufficient worthwhile openings back in Australia to entice them home. Another essential, as studies over twenty-five years at the University of Syracuse, New York, have shown, is that it is vital for the 'mother' university or institution to stay in touch with the traveller, giving news and assurance that their achievements are being noticed. Then there is much more likelihood that the brain will be repatriated and not permanently drained away. Australians like to come home. On the other hand: they don't like to be humiliated. They deserve a decent job.

On government funding: it is clear that Australia is going against an international trend. We spend $3.75 billion of Commonwealth money on scientific research. This is not matched as it might be by funds from the private sector (whose contribution is plummeting). I suggested to the prime minister and some of his colleagues (at the PM's Science Council in December 1999) that, if they were concerned by the voters' possible unwillingness to countenance a greater expenditure they may like to bear in mind that Australians splurge over $105 billion per annum on gambling, a spending spree that has clearly become pathological! There is also useful guidance to the amount a nation might allocate effectively to R & D. Sir Robert May has published several papers in the journal *Science* relating what he calls 'the bang for the buck' — the degree of fiscal tightness that nonetheless produces good results. Britain seems to be in that range. Australia is below it. No blank cheques are being demanded. Effective funding is between $1 billion and $3 billion short of what it should be.

On the recruitment of youth: it is important to remember that we are not worrying about a shortage of scientists. We should be worrying instead about a failure to prepare the next generation for a world dominated by science and technology. In 1951 there were 32 000 students enrolled in university in Australia. By 1997 the figure was 660 000. So, even with a fall in recruitment, we are still producing more science graduates than ever before.

The real concern behind the figures is that Australians are less informed about scientific matters than they should be, far less. Talk about a 'Knowledge Society' or the 'Clever Country' amounts to so much sloganeering unless backed by the kind of resources a mature society should be able to afford. As for the 'sizzle and meat' demanded by *The Australian* newspaper for our schools, it is worth remembering that it requires a flow of trained, enthusiastic science teachers to achieve a warmed-over snack. At a teachers' conference in New South Wales in 1999, I was told that only seventeen mathematics teachers had graduated in that state the year before. A fellow of the Academy of Science added, 'I expect the number for physics will be two!' Good figures are difficult to come by but, whatever they are, it is clear that the reality is disastrously below what is required. Furthermore, the very sensible (even vital) preparation needed in primary schools has not been possible because of the same trained-teacher shortage. Professor Don Watts, a former Vice Chancellor and now Professor of Education at Notre Dame University in Perth, claims that political rhetoric has set up the primary teachers to fail: there is much expectation, no backing.

There is little mystery about the nature of the sizzle itself. Old-fashioned science courses (such as those I endured) are 95 per cent tedium, especially if they seem directed at an unending preparation for some professional qualification you know, in the end, you don't want. The hard grunt of rote learning can now be done outside the classroom using interactive computer technology. The pace is thus set by the student and will not affect the class by bringing it down to the progress of the slowest. The practical lessons, as the Taiwanese have shown, can be directed towards problem-solving and the use of a wide range of disciplinary skills in the process. This leaves plenty of time for reading and the proper use of the classroom: for the discussion of *IDEAS*.

The next task (as discussed in chapter 2) is to blast science out of its ghetto. It should be integrated into every subject, related explicitly to every possible job. If that is too difficult to imagine, here is a list: Religious Studies (add: archaeology, psychology); English (linguistics, science fiction, computerised analysis of style); Sport (health, psychology, physics), History (scientific standards of evidence, sociology); Music (physics, engineering, psychology); and so on.

Do the employers of the next generation of Australian youth agree there is a crisis? That they do seemed clear at the Innovations Summit, as *New Scientist* reported:

> University scientists weren't the only ones who were outspoken about the damage being done to the system's research capacity by budget cuts. They were firmly supported by business leaders. Chris Knoblanche, the Australian head of consulting firm Arthur Anderson, told the meeting that science is 'losing the battle for talent'. As a business consultant, he lamented the fact that too many bright young people were heading into business rather than research careers. 'Before long business will have nothing to sell'.

There will, of course, be plenty to sell. But it will be foreign goods. The implication, however, is that our youth will be equipped only as service personnel or as junior managers, mere cadres of underlings for the behemoths from abroad using Australia as service base and shopping mall. The genial Australian buffoon of yesteryear will have returned, the reincarnation of Bazza McKenzie from the 1960s will be Darren or Kylie of 2005 genuflecting to the requirements of distant masters.

Such is the present predicament. It is not obvious to the general Australian community because it takes a generation for the real signs of neglect to become apparent. Just as research itself is slow to show results, so harm done to the scientific infrastructure will not be apparent until it is too late — 2005 or later. The consequences of the damage could be profound. Australia would become a second-rate society and largely unable to determine its own destiny. The malaise made manifest by Pauline Hanson and her party in the 1990s would be displayed in various other forms, more stridently. The standard of living, especially that measured by GPI, would go down substantially. By then bewildered Australians would wonder what went wrong. But there is an alternative.

Why not a Clever Country?

It was not a coincidence that both the Innovations Statement (1995, Keating) and the Innovations Summit (2000, Howard) took place in the same place with the same rhetoric. Politicians in Australia are constantly rediscovering science, falling in love with it, then wandering back to normal life as if nothing had happened. Barry Jones noted that ministers for science here tend to be on the way up or the way down (he himself was parked). They bring nothing to the portfolio (exceptions: Jones, McGauran, Free, possibly Crean) and say so unblushingly. Senator Minchin told a squirming Australia Prize banquet that his own science master at Knox would not believe Nick had taken the reins of science for the nation, so abysmal was his schoolboy accomplishment. The embarrassment was not only his willingness to display ignorance, it was that so many ministers before him had done likewise. In fact the prime minister said exactly the same about himself directly afterwards.

The exceptions show how straightforward it is to get a proper grounding. When appointed shadow to Barry Jones during the epic stretch of the Hawke/Keating administration, Peter McGauran asked Jones, collegially, how he could best prepare himself for his unexpected and unfamiliar responsibilities. Jones advised him to get on the road and look at what Australian science consists of and to listen. Listen critically. As a result of this thorough induction, Peter McGauran became an excellent minister for science, no matter that he sometimes resembled a cross between Tigger and Ginger Meggs. He had done his homework and was universally respected by the scientific establishment. Not because he gave them what they wanted — he was not allowed to — but because he showed he understood the issues.

High order dedication may not be open to many science ministers (whose average tenure is eighteen months!), particularly if they are carrying the immense Industry portfolio in tandem. But without some effort from those several ministers responsible for the carriage of scientific affairs in general (Education, Industry, Communications, Environment, Trade) it is unlikely that science and technology will be taken at all seriously except as an add-on. It is all very well for politicians, notably John Howard, to say how much they like their encounters with science but, unless results are forthcoming, it will be seen only as a six-monthly day off among pleasant supplicants.

It is not a matter of ideology either. William Waldegrave, Margaret Thatcher's minister, had a formidable grasp of scientific affairs and quoted from the latest books, which he had obviously read. Malcolm Fraser knew about CSIRO history and the organisation's relationship to industry and agriculture; he also took a disconcertingly (to his ministers) intense interest in details of environmental science. Neville Wran, as Premier of New South Wales, gave himself an intensive education while on the job. Peter Beattie in Queensland appears to be following suit.

But it is at the national level that erudition is required, across cabinet and the opposition front bench. Two aspects of science policy have to be understood: 1. how much it is linked (like education, above) to every portfolio; and 2. that funding from private enterprise is a long way off and Commonwealth backing is needed, at least for the time being. No nation in modern times can get away with forcing its scientists to find alternative funding by sending them out shoeless with a begging bowl.

Australia has some spectacular advantages which will endure to give it an excellent chance for recovery. They are:

1   Our geography. We are close to and trusted by many nations needing scientific co-operation few others could supply. Indonesia, Papua New Guinea, South-East Asia all have village-based societies needing the kind of technology and agriculture at which Australia is expert. We also have the gift of time-zone — we are in easy communication with the largest populations in the world.

    We are so large that astronomers, marine scientists and meteorologists across the globe must deal with us. Or set up their own bases! The same is true of our huge allocation of the Antarctic continent.

2   We have a brilliant scientific tradition. While it can be an embarrassing sign of immaturity (as *New Scientist*'s Ian Anderson has pointed out) to be claiming 'world class' all the time, it is true that the achievements to date have been

considerable. They are certainly enough to be the basis of a rebound.

3   Our young people are interested. In chapter 2 I suggested there was a gap between stated public interest in science and reality. Young people may be stampeding away from traditional science-in-a-box, but they are nonetheless very keen on wildlife, environment, computers, health and any science that makes them rich.

4   There is a science-awareness infrastructure in place, however fragile, that can be built up by adding compost (resources). The plant still lives.

5   Australia, at times, may be disconcertingly inept and inert, but it is equally adept at becoming galvanised and intrepid. This may be the characteristic of a young society with a small, ever-changing population. But, as America and Israel have shown, it is ideal for a scientific culture.

6   There is no longer a need, in the new global village, to be transplanted to the heavy-duty centres in Western Europe or America. Your enterprise can sit serenely under the coolabah tree with its rainbow lorikeets as you communicate in comfort with the rest of the world, inventing the future.

Australian society is likely to understand quickly, if given the chance, that there are at least five good reasons for being good at science. Once the intellectual revolution is underway, they will be self-reinforcing. The intellectual revolution needs to happen only once.

CONCLUSION

When Bill Gates was setting up his second research base outside Seattle (he chose the ancient city of Cambridge) he issued only three riding instructions:

1 Hire the best people you can find and let them do what they want.
2 If all your projects succeed, you have failed.
3 Remember you are European, not simply British: hire from far and wide.

This brief matches that of my hero from another age, Max Perutz, who ran the renowned Cambridge Microbiology Lab with such brilliance a generation ago. It stands as a yardstick for R & D in Australia as we enter the next phase of history. We compare very badly:

1 Only a few favoured outfits can afford to build up an A-team and allow it to roam free. Most Australian scientists (in universities anyway) are grimly chained to the treadmill of writing grant applications for the few dollars available. The pity of this is that mediocrity ends up costing more because excellence creates its own momentum and then the second-rate fall away; the first-rate also produce ideas that end up paying for themselves in the long run.
2 If all your projects succeed, then you've obviously been playing safe. Even Senator Richard Alston has noticed that the secret of Silicon Valley and the IT revolution was that so many people were prepared to face failure, then try again. Australia imagines it cannot afford to take a scientific chance. This caution stands in bewildering contrast to its obsession with gambling. But then gambling is a very short-term occupation.
3 We do recruit scientists from our region. But we are not sufficiently engaged in it. 'Me too' science that is enmeshed in the traditional markets of Europe and America will see us but as

an adjunct to the giants. Proper links to Asia and the Pacific will confirm our role as a unique force.

Why is there a disjunction between what we could be doing and our present slump? I would suggest it is because Australians have little idea where they are headed and are offered no leadership at the political level about possibilities. In fact we spend more time floundering in historical crises (monarchy, stolen generation) than in contemplating possible futures. A nation which is uncertain about its past has little hope of confidently moving forward.

Barry Jones, Minister for Science in the Hawke ministry, set up the Commission for the Future in the mid-1980s. Its brief was to foster debate on where the country was going and what part science and technology would play in its journey. The Commission was one of the first bodies to be eliminated when the Howard government came to office in 1996. This was a pity: it could have come into its own just at a time when the nation seems bereft.

What kind of advice could the Commission have offered? Well, not advice as such: like science itself, the Commission was never instructive. Instead it offered thought experiments. Here is one possibility:

Sir Peter Hall, writing about the growth and decline of cities, adapted Schumpeter and saw technological revolutions occurring every fifty years or so, with communications linked to transport providing the impetus. Thus, 150 years ago, with the railways, came the telegraph, the typewriter and the postage stamp; 100 years ago, when cars arrived, so did radio, film and the telephone; 50 years ago it was jet planes, television and the mainframe computer. Now we have the Internet, laptops — and gridlock. Four years ago traffic jams cost the United States US$170 billion in a year!

So it could be that we have reached a new era in which having to move is an impost, staying put a benefit? How could you design a community in Australia for 2020 to suit such a transformation? Businesses and research teams could work anywhere, there would be no tyranny of distance. The key would be in identifying which fields would suit our region most, how to recruit the best and the brightest and ways to link them to the communities where the problems need to be solved.

The Landcare movement, begun in 1990 by Joan Kirner (then a minister in the Victorian Government), Rick Farley (of the National Farmers' Federation) and Phil Toyne (of the Australian Conservation Foundation) is an example of what can be achieved. Kirner, who was to become, briefly, Victorian premier, found that everyone had policies for the (overwhelmingly scientific) challenges of blighted forests, ruined soils and worried people. Few had bothered to go to the places affected, talk to the locals or look at the evidence. Landcare put old enemies on the same side, linked disparate communities and put them all in touch with useful science. So far so good. But the problems remain huge. When seen, however, in the context of the future of Australia and its neighbours, and combined with a dialogue using communications technology effectively, there could be a new momentum. It is badly needed. If it fails, Australia could actually become a net importer of food!

Landcare has shown how science can become integrated into a broad community approach to a set of problems. It is a rare example. Australia is very good at trying new ideas but very bad at sticking with them. Those of us who accuse this country of mediocrity are often asked which nation is better. That is not the point. Australia is its own best comparison. When it sets high standards they can be second to none. When it courts mediocrity (as Donald Horne and Phillip Adams have documented) it can be dire.

There are compelling signs that Australian science is moving away from the edge of excellence, both as a separate pursuit and certainly as a publicly integrated activity. The neglect is particularly worrying when one realises how magnificent the benefits could be.